Benoît Delahaut

Option Pricing Models built from Lévy Processes

Benoît Delahaut

Option Pricing Models built from Lévy Processes

An Empirical Comparison

Social Sciences Series

Impressum / Imprint

Bibliografische Information der Deutschen Nationalbibliothek: Die Deutsche Nationalbibliothek verzeichnet diese Publikation in der Deutschen Nationalbibliografie; detaillierte bibliografische Daten sind im Internet über http://dnb.d-nb.de abrufbar.

Alle in diesem Buch genannten Marken und Produktnamen unterliegen warenzeichen-, marken- oder patentrechtlichem Schutz bzw. sind Warenzeichen oder eingetragene Warenzeichen der jeweiligen Inhaber. Die Wiedergabe von Marken, Produktnamen, Gebrauchsnamen, Handelsnamen, Warenbezeichnungen u.s.w. in diesem Werk berechtigt auch ohne besondere Kennzeichnung nicht zu der Annahme, dass solche Namen im Sinne der Warenzeichen- und Markenschutzgesetzgebung als frei zu betrachten wären und daher von jedermann benutzt werden dürften.

Bibliographic information published by the Deutsche Nationalbibliothek: The Deutsche Nationalbibliothek lists this publication in the Deutsche Nationalbibliografie; detailed bibliographic data are available in the Internet at http://dnb.d-nb.de.

Any brand names and product names mentioned in this book are subject to trademark, brand or patent protection and are trademarks or registered trademarks of their respective holders. The use of brand names, product names, common names, trade names, product descriptions etc. even without a particular marking in this works is in no way to be construed to mean that such names may be regarded as unrestricted in respect of trademark and brand protection legislation and could thus be used by anyone.

Coverbild / Cover image: www.ingimage.com

Verlag / Publisher:
AV Akademikerverlag
ist ein Imprint der / is a trademark of
OmniScriptum GmbH & Co. KG
Heinrich-Böcking-Str. 6-8, 66121 Saarbrücken, Deutschland / Germany
Email: info@akademikerverlag.de

Herstellung: siehe letzte Seite /
Printed at: see last page
ISBN: 978-3-639-64081-6

Contents

List of Figures

List of Tables

Introduction

Options are financial instruments that give their owners a right and its counter-party a duty. Whilst the first listed options appeared on the Chicago Board of Exchange in 1973 and became more actively traded, it is said that the use of options dates back to 332 BC when Greek philosopher Thales bought the right to buy olives before the harvest. What is amusing is that the values quantify-ing several different risks related to an option are now called the Greeks. Since their creation, the main problem that arose with options can be summed-up in the sentence: "How to fairly price them?". Until now, in option pricing, the most frequent reference is Black-Scholes which refers to the model introduced in the paper published by F. Black and M. Scholes forty years ago in Black and Scholes (1973), the year in which options started trading on the CBOE. The reason is its simplicity - the model is easy to both understand and implement : it only depends on one parameter which is $\hat{\sigma}$, the implied volatility.

But since 1973, the model has been criticised quite a lot, as it suffers major drawbacks. Even though it is interesting from a theoretical point of view, the assumptions are simply too strong to fit the real-world data. Interest rates are not constant, there are transaction costs, markets are not perfectly liquid... All these observations already violate the B-S assumptions, but even worse, the distribution of the stock price, assumed to follow a geometric Brownian motion with constant drift and volatility according to B-S, shows other features on the market. In fact, the distribution of the returns is leptokurtic, i.e. has a higher kurtosis than the

1

normal distribution, which means that the empirical distribution of the returns has fat tails. As a consequence, highly improbable events, such as the 2008 crisis, are more probable than predicted by the B-S model. It also implies the existence of a volatility smile : at-the-money options' volatility is lower than in-the-money or out-of-the-money options' volatility where the B-S model expects same volatility across the whole range of moneyness. Also, it has been showed that the distribution of the returns does not have a zero skewness but tends to show a negative one. In other words, large losses are more likely to happen than large gains. The phenomenon of volatility clustering, stating that a highly volatile period should remain volatile as a quiet one should remain quiet is not explained by B-S. These facts point out important drawbacks of the B-S model, and more elaborated models, using local volatility or stochastic volatility, have been introduced. The other big issue is that in practice, stock price processes are discontinuous, and this is where Lévy processes, also called jump processes, come into play.

In this thesis, we study two specific exponential Lévy processes, the VG and the Merton processes, and apply two different option pricing methods although the method holds for more general processes. We start with a quick explanation of the background of Lévy processes and Fourier transforms, and motivate our definitions. In chapter 3, we explain in details the pricing method using the fast Fourier transform for models whose characteristic function of the log stock process is known, introduced in Carr and Madan (1999). In the following section, we briefly explain another pricing method based on the cosine series expansion of the risk-neutral density function's Fourier transform. After that, we introduce the Variance Gamma model and the Merton models that will be used to test the pricing methods. Finally, we compare the different methods, discuss the choice of the parameters and calibrate them to market data, some S&P 500 European options with mid- to long-term maturities. We also tackle the calibration problem using two different "weighted least squares functions", one aiming at approximating the option prices, the other one focusing on the approximation of the implied volatility, which is useful in practice.

Chapter 1

Definition and basic properties of Lévy processes

This class of processes, also called jump processes, were named after the French probabilist Paul Lévy. Apart from introducing them, Lévy is known as being one of the founders of the probability theory and developed useful tools even though he was known not to be a "formalist". In the following we assume that the very base of probability theory is known. For more details about it, see Shreve (2004).

1.1 Lévy processes

Definition 1.1 (càdlàg processes). Let $(\Omega, \mathcal{F}, \mathbb{P})$ be a probability space. A process $X = (X_t)_{t \in [0,T]}$ with $T \in \bar{\mathbb{R}}_+$ is called a càdlàg process (for *continu à droite* and *limites à gauche* in french) if it almost surely has sample paths that are right continuous with left limits i.e. for almost all $\omega \in \Omega$ the following holds :

1. $\lim_{s \to t_+} X_s(\omega) = X_t(\omega) \; \forall t \in [0, T[$,

2. $\lim_{s \to t_-} X_s(\omega) = X_{t-}(\omega)$ exists $\forall t \in \,]0, T]$.

This property is the one that allows for jumps (i.e. discountinuity over time). A legitimate question would be : "Should we consider càdlàg or càglàd (left continuous with right limits) processes?". On a strict theoretical point of view, there is no essential difference between both kinds of processes, but the point is that in practice, time passes and does not revert. As a jump at time t needs to appear as an unforseeable event, the right continuity is appropriate as one could predict the value of $X_t(\omega)$ following the path along t for ω fixed if it were left continuous.

Definition 1.2 (Lévy process). Let $(\Omega, \mathcal{F}, \mathbb{P})$ be a probability space. Then we define a Lévy process on this space as a stochastic process $X = (X_t)_{t \geq 0}$ with value in $\mathbb{R}^d$, $d \in \mathbb{N}_0$ such that

1. X is a càdlàg process,

2. $\mathbb{P}(X_0 = 0) = 1$,

3. For $0 \leq s \leq t$, $X_t - X_s$ is equal in distribution to X_{t-s} (stationary increments),

4. For $0 \leq s \leq t$, $X_t - X_s$ is independent of $\{X_u, u \leq s\}$ (independent increments).

Let us notice from the definition that a Lévy process is not very different from a Brownian motion: both processes have independent and stationary increments, but a Lévy process loses the normal distribution of the increments. As the requirements for being a Brownian motion are stronger than those for being a Lévy process, a Brownian motion is a special case of Lévy process.

Although a general definition considers Lévy processes with values in $\mathbb{R}^d$, we will consider the special case where $d = 1$.

1.2 Poisson processes

The main innovation of the Lévy processes is the presence of jumps. To understand what a Lévy process could look like, it is worth explaining the basic jump processes

: the Poisson and the compound Poisson process. First of all, let us remember what is an exponential random variable X.

Definition 1.3 (Exponential random variable). A continuous r.v. X is an exponential r.v. with parameter λ if it has the following density function :

$$f_X(x) = \lambda e^{-\lambda x} \mathbf{1}_{x \geq 0}.$$

If U is uniformly distributed on $[0, 1]$, $-\frac{1}{\lambda} \ln(U)$ is exponentially distributed with parameter λ (useful for computational purposes). The exponential random variable has the interesting "absence of memory" property and is characterised by it :

Proposition 1.4. *Let $T \geq 0$ be a nonzero random variable such that*

$$\forall t, s > 0, \mathbb{P}(T > t + s | T > t) = \mathbb{P}(T > s),$$

then T is a an exponential random variable.

The proof follows from Bayes rule and is given in Cont and Tankov (2004), chapter 2. Let us notice that we chose the letter T as the random variable can be interpreted as a random time.

Definition 1.5 (Poisson random variable). A discrete r.v. X is a Poisson r.v. with parameter λ if it has the following density function :

$$\forall n \in \mathbb{N}, \mathbb{P}(X = n) = e^{-\lambda} \frac{\lambda^n}{n!}.$$

We are now ready to link the exponential r.v. to the Poisson r.v. via the definition of the Poisson process.

Definition 1.6 (Poisson process). Let $(\tau_i)_{i \in \mathbb{N}}$ be a sequence of i.i.d. exponential r.v. with common parameter λ and let $T_n = \sum_1^n \tau_i$. The process $(N_t)_{t \geq 0}$ defined by

$$N_t = \sum_{n=1}^{\infty} \mathbf{1}_{t \geq T_n}$$

is called a Poisson process with intensity λ.

By construction, a Poisson process is a Lévy process (the simplest non trivial Lévy process with jumps) and the paths $t \to N_t(\omega)$ are almost surely piecewise constant, the distance between two jumps being given by an exponential random variable of parameter λ. Its name is due to the fact that $\forall t \in \mathbb{R}_0^+$, N_t follows a Poisson distribution with parameter λt :

$$\forall n \in \mathbb{N}, \mathbb{P}(N_t = n) = e^{-\lambda t} \frac{(\lambda t)^n}{n!}.$$

Obviously the Poisson process in itself is not very useful in practice as it only allows for jumps of size 1. More realistic jumps can be achieved by a first generalisation of the Poisson process: the compound Poisson process.

Definition 1.7 (Compound Poisson process). Let $(N_t)_{t \geq 0}$ be a Poisson process with intensity λ, then a compound Poisson process with intensity λ and jump size distribution f is a stochastic process X_t defined as

$$X_t = \sum_{i=1}^{N_t} Y_i,$$

where the random variables Y_i are i.i.d. with density f and the r.v. N_t is independent from Y_i $\forall i$.

It means that a compound Poisson process is a Poisson process where the jumps, instead of being of size 1, are i.i.d. from a r.v. with a known density function f. A compound Poisson process is a Lévy process and it can be shown that any Lévy process with piecewise constant sample paths is a compound Poisson process.

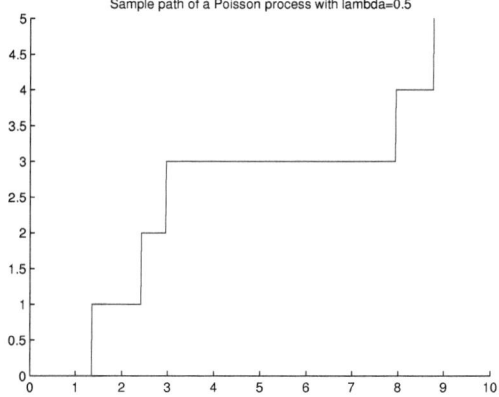

Figure 1.1: Poisson process

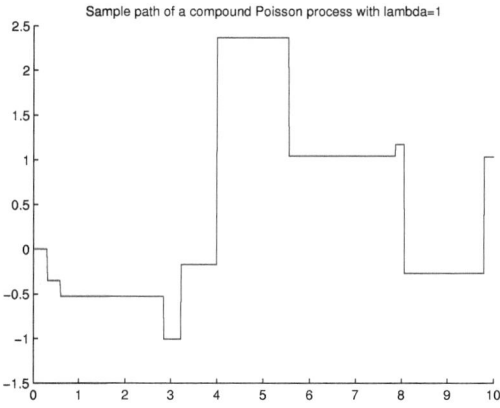

Figure 1.2: Compound Poisson process

8

1.3 More properties of Lévy processes

The Lévy measure is one of the fundamental concepts introduced via the Lévy process and serves for its characterisation.

Definition 1.8 (Lévy measure). Let $(X_t)_{t \geq 0}$ be a Lévy process on $\mathbb{R}^d$. We define the Lévy measure ν on $\mathbb{R}^d$ by

$$\nu(A) = \mathrm{E}\big(\#(t \in [0,1] : \Delta X_t \neq 0, \Delta X_t \in A)\big), \quad A \in \mathcal{B}(\mathbb{R}^d),$$

i.e. the number of expected jumps whose size belongs to A.

We are now in possession of all the ingredients necessary to formulate two fundamental results about Lévy processes. The first one is a consequence of the Lévy-Itô decomposition, not given here, stating that each Lévy process (X_t) is entirely determined by a characterstic triplet (A, ν, γ), where A is a positive definite matrix $d \times d$, ν is a positive measure and γ is a vector of length d. A and γ determine the Brownian motion component of the Lévy process, and ν determines the jump component. This triplet is called the characteristic triplet of the process (X_t). Let us mention here that in the case of a 1-dimensional Lévy process, our case, the matrix A is reduced to the variance of the Brownian motion and the γ is its drift.

The second important result here is called the Lévy-Khintchine formula. It gives a formula expressing the characteristic function (see next chapter) of a Lévy process in terms of its characteristic triplet (A, ν, γ). It is of paramount importance in our work as this formula is used in practice to compute the characteristic functions of Lévy processes. Here we give the one-dimensional version.

Theorem 1.9 (Lévy-Khintchine formula). *Let $(X_t)_{t \geq 0}$ be a Lévy process on $\mathbb{R}$ with characteristic triplet (A, ν, γ). Then*

$$E(e^{izX_t}) = e^{t\psi(z)}$$

9

with

$$\psi(z) = -\frac{1}{2}Az^2 + i\gamma z + \int_{-\infty}^{\infty} (e^{izx} - 1 - izx\mathbf{1}_{|x|\leq 1})\nu(dx).$$

Chapter 2

Fourier transform and characteristic function

The processes that we use in our work share one mandatory common property in that the characteristic functions of their risk-neutral densities are analytically known. This is by far less restrictive than requiring the density function to be known, as there are plenty of models whose density function is not analytically known but whose characteristic function is.

We start this chapter with the definition of the Fourier transform of a given function. Such a function needs to fulfil "some" hypotheses that are not described here, and is called a *reasonable* function. It is enough to know that a density probability function is *reasonable*.

Definition 2.1 (Fourier transform). Let $f : \mathbb{R} \to \mathbb{R}$ be a *reasonable* function, its Fourier transform $\hat{f}$ is defined as

$$\hat{f}(u) = \int_{-\infty}^{\infty} e^{iux} f(x) \mathrm{d}x.$$

We also use the notation $\hat{f} = \mathcal{F}(f)$, with $\mathcal{F}$ being the Fourier transform operator.

One of the most important properties of the Fourier transform is that the knowledge of it allows to recover the original function (in our case the density function) through the use of the inverse Fourier transform formula:

$$f(x) = \frac{1}{2\pi} \int_{-\infty}^{\infty} e^{-iux} \hat{f}(x) \mathrm{d}u.$$

We also write : $f = \mathcal{F}^{-1}(\hat{f})$ or $f = \mathcal{F}^{-1}(\mathcal{F}(f))$. In literature, other equivalent definitions can be found but we will stick to this one.

Definition 2.2 (Characteristic function of a random variable). Let X be a (real-valued) r.v. whose density function f_X exists. We define its characteristic function Φ_X for $u \in \mathbb{R}$ by

$$\Phi_X(u) = \mathrm{E}(e^{iuX}) = \int_{-\infty}^{\infty} e^{iux} f_X(x) \mathrm{d}x.$$

In other words, the characteristic function $\mathcal{F}(f_X)$ of a random variable X is simply the Fourier transform of its density function, f_X, and this density function is "characterised" by its characteristic function, hence its name. The characteristic function of a density function is well-defined because for $x \in \mathbb{R}$, $e^{ix} = \cos(x) + i\sin(x)$ so that $|e^{ix}| = 1$. As a consequence, for f_X a density function and Φ_X its characteristic function, we have the following relationship :

$$f_X(x) = \frac{1}{2\pi} \int_{-\infty}^{\infty} e^{-iux} \Phi_X(u) \mathrm{d}u.$$

The use of the Fourier transforms, or characteristic functions, is often very convenient as they have nice properties. We first define the convolution product and then give some of those properties. For a more exhaustive list of them see Schmelzle (2010), p. 9.

Definition 2.3 (Convolution product). Let f, g be functions defined on $\mathbb{R}$. We define the convolution product of f and g as

$$(f * g)(x) = \int_{-\infty}^{\infty} f(x - y) g(y) \mathrm{d}y$$

Proposition 2.4. *We still denote by $\mathcal{F}$ the Fourier transform operator. Let $a, b \in \mathbb{R}$, $n \in \mathbb{N}_0$ and f, g be reasonable functions, then we have the following properties :*

12

1. $\mathcal{F}(af + bg) = a\mathcal{F}(f) + b\mathcal{F}(g)$ *(linearity of the operator)*,

2. *in case, $f^{(n)}$ is piecwise continuously differentiable with its derivatives absolutely integrable on $\mathbb{R}$, then*

$$\mathcal{F}\big(f^{(n)}\big)(u) = (-iu)^n \mathcal{F}(f)(u),$$

3. $\mathcal{F}(f * g) = \mathcal{F}(f)\mathcal{F}(g)$ *and* $\mathcal{F}(fg) = \mathcal{F}(f) * \mathcal{F}(g)$,

Needless to say that transposing in the Fourier space often highly simplifies calculations. This makes Fourier transforms very useful in practice - in solving the partial differential equations of the Heston model, for example.

Chapter 3

Two option pricing methods

In this chapter we observe and compare two popular option pricing methods. In fact the first method, introduced in Carr and Madan (1999), is not really used anymore *per se* but has formed the base of many others. The second method dates back to 2008 and exploits the cosine expansion of the inverse Fourier integral to compute option prices. Let us notice that unlike the CM method, the *cos* method does not require the FFT algorithm.

The starting point is that a european call option can be expressed as the expected value of its future payoff discounted according to the interest rates. It means that if $\mathbb{Q}$ is the risk-neutral measure of the stock price, we have the following :

$$C_T = e^{-rT} \mathrm{E}^{\mathbb{Q}}\big((S_T - K)^+ | \mathcal{F}_0\big), \tag{3.1}$$

T being the time from now to maturity, S_T the stock price at expiration, K the strike price and $\mathcal{F}_0$ representing the information available at $t = 0$. This formula is a general formula and is valid for every model. The only assumption it makes is the no-arbitrage hypothesis.

We now focus on the pricing of a European call option maturing at T and with strike price K. We denote by s_t the neperian logarithm of S_T : $s_t = \ln(S_T)$.

Consequently, the characteristic function of s_T is given by :

$$\Phi_T(u) = \mathrm{E}(e^{ius_T}).$$

In many cases this characteristic function is known. It includes all the cases where the log price follows an infinitely divisible process with independent increments for example. Such models are the VG process, the inverse Gaussian law, the Heston model,... If the characteristic function is analytically known, the risk-neutral probability of finishing in-the-money is (see Bakshi and Madan (2000))

$$P(S_T > K) := \Pi_2 = \frac{1}{2} + \frac{1}{\pi} \int_0^{\infty} \Re\left(\frac{e^{-iu\ln(K)}\Phi_T(u)}{iu}\right) du,$$

and the delta of the option is given by

$$\Pi_1 = \frac{1}{2} + \frac{1}{\pi} \int_0^{\infty} \Re\left(\frac{e^{-iu\ln(K)}\Phi_T(u-i)}{iu\Phi_T(-i)}\right) du.$$

If there we assume no dividend and a constant continuous risk-free rate r, the price of the option is determined by :

$$C = S\Pi_1 - Ke^{-rT}\Pi_2$$

As the integrand is singular at $u = 0$, this formula cannot be used in this current form to find the option price.

3.1 The Carr-Madan method

In 1999, Carr and Madan introduced the following method using the fast-Fourier transform, an efficient algorithm to compute the discrete Fourier transform which approximates the Fourier transform. We now explain in details how this method works as it has inspired various other ones, following the schema given in Carr and Madan (1999).

3.1.1 The explicit formula for a call price

Let $k = \ln(K)$ and let $C_T(k)$ be the call price at $t = 0$ of an option maturing at T with strike $K = e^k$. Furthermore we assume that the log price at expiration s_T has the risk-neutral density q_T, which might be unknown. The characteristic function of s_T is by definition

$$\Phi_T(u) = \int_{-\infty}^{\infty} e^{ius} q_T(s) ds.$$

By applying 3.1 $C_T(k)$ is given by

$$C_T(k) = \int_{k}^{\infty} e^{-rT}(e^s - e^k) q_T(s) ds$$

As $C_T(k) \to S_0$ when $k \to -\infty$, $C_T(k) \notin \mathcal{L}^2(\mathbb{R})$, the Hilbert space of square-integrable functions, which is a problem when considering the Fourier transform of the call price. To avoid this, we will consider a modified call price $c_T(k)$ that belong to $\mathcal{L}^2(\mathbb{R})$:

$$c_T(k) = e^{\alpha k} C_T(k), \ \alpha > 0.$$

The function $e^{\alpha k}$ is one of the numerous functions that could be used here to make a modified call belong to $\mathcal{L}^2(\mathbb{R})$. We will now compute the Fourier transform of the modified call and then its inverse transform. This manipulation, though apparently useless, enables us to express the call price in terms of the characteristic function. The Fourier transform of $c_T(k)$ is given by :

$$\psi_T(v) = \int_{-\infty}^{\infty} e^{ivk} c_T(k) dk,$$

and by performing the inverse Fourier transform formula, we get that

$$C_T(k) = \frac{e^{-\alpha k}}{2\pi} \int_{-\infty}^{\infty} e^{-ivk} \psi_T(v) dv = \frac{e^{-\alpha k}}{\pi} \int_{0}^{\infty} e^{-ivk} \psi_T(v) dv \qquad (3.2)$$

because $C_T(k)$ being real, ψ_T is odd in its imaginary part and even in its real part. We furthermore notice that $\psi_T(c)$ can be expressed in terms of the characteristic function Φ_T as :

$$\psi_T(v) = \int_{-\infty}^{\infty} e^{ivk} \int_k^{\infty} e^{\alpha k} e^{-rT} (e^s - e^k) q_T(s) ds d0k$$

$$= \int_{-\infty}^{\infty} e^{-rT} q_T(s) \int_{-\infty}^s (e^{s+\alpha k} - e^{(1+\alpha)k}) e^{ivk} dk ds$$

$$= \int_{-\infty}^{\infty} e^{-rT} q_T(s) \left(\frac{e^{(\alpha+1+iv)s}}{\alpha + iv} - \frac{e^{(\alpha+1+iv)s}}{\alpha + 1 + iv} \right) ds$$

$$= \frac{e^{-rT} \Phi_T(v - (\alpha+1)i)}{\alpha^2 + \alpha - v^2 + i(2\alpha + 1)v}. \tag{3.3}$$

Inserting now 3.3 in 3.2, we obtain the following expression for the call price:

$$C_T(k) = \frac{e^{-\alpha k}}{\pi} \int_0^{\infty} e^{-ivk} \frac{e^{-rT} \Phi_T(v - (\alpha+1)i)}{\alpha^2 + \alpha - v^2 + i(2\alpha + 1)v} dv. \tag{3.4}$$

For $\psi_T(v)$ to exist, the modified call needs to be integrable. The modifying factor $e^{\alpha k}$ "helps" the integrand in the negative values of k but worsens it when $k \to +\infty$. The condition $\psi_T(0) < +\infty$ or equivalently $\Phi_T(-(\alpha+1)i) < +\infty$ ensures that the modified call is integrable. This can be written as

$$\Phi_T(-(\alpha+1)i) = \int_{-\infty}^{\infty} e^{(\alpha+1)s} q_T(s) ds = \int_{-\infty}^{\infty} S^{\alpha+1} q_T(s) ds = E(S_T^{\alpha+1}) < +\infty.$$

We now want to have an estimate of the truncation error of the integral 3.2. As $v \in \mathbb{R}$ we get that $|\Phi_T(v - (\alpha+1)i)| \leq E(S_T^{\alpha+1})$. As $|a||b| = |ab|$ and $|c|^2 + |d|^2 = |c + di|^2 \; \forall a, b \in \mathbb{C}, \; \forall c, d \in \mathbb{R}$ and because $r, T \in \mathbb{R}^+$, using 3.3 it can be shown that

$$|\psi_T(v)|^2 \leq \frac{E(S^{\alpha+1})}{(\alpha^2 + \alpha - v^2)^2 + (2\alpha + 1)^2 v^2} \leq \frac{A}{v^4}, \tag{3.5}$$

where the second inequality is due to the fact that $v^4 \geq (\alpha^2 + \alpha - v^2)^2 + (2\alpha + 1)^2 v^2$ and where $A = E(S_T^{\alpha+1})$. From 3.5 we deduce that

$$|\psi_T(v)| < \frac{\sqrt{A}}{v^2},$$

and the upper tail of the integral 3.2 is bounded by

$$\int_a^{\infty} |\psi_T(v)| dv < \frac{\sqrt{A}}{a}.$$

17

It then follows that the truncation error in the former equation is bounded by

$$\frac{e^{-\alpha k}}{\pi} \frac{\sqrt{A}}{a}.$$

3.1.2 Use of the fast Fourier transform

We are left with the computation of 3.2, which is an exact application of the Fast Fourier Transform algorithm, an efficient algorithm that computes expressions of the following kind :

$$X(k) = \sum_{j=1}^{N} \omega_N^{(j-1)(k-1)} x(j), \quad k = 1, ..., N. \tag{3.6}$$

Here ω_N denotes the Nth root of the unit $e^{-\frac{2\pi i}{N}}$. One restriction of this algorithm is that it needs N to be a power of 2. To have a formula of the form 3.6 we first approximate the integral in 3.2 by the following formula using the trapezoid rule :

$$C_T(k) = \frac{e^{-\alpha k}}{\pi} \sum_{j=1}^{N} e^{-iv_j k} \psi_T(v_j) \eta, \tag{3.7}$$

where $\eta \approx dv$, $v_j = \eta(j-1)$ and the truncation of the integral is $a = N\eta$. If we further define $\lambda = \frac{2\pi}{\eta N}$, $b = \frac{1}{2}N\lambda$ and $k_u = -b + \lambda(u-1)$, $u = 1, ..., N$, we get after simplifications

$$C(k_u) = \frac{e^{-\alpha k_u}}{\pi} \sum_{j-1}^{N} \omega_N^{(j-1)(u-1)} e^{ibv_j} \psi(v_j) \frac{\eta}{3} \big(3 + (-1)^j - \delta_{j-1}\big), \tag{3.8}$$

δ_j being the Kronecker symbol, i.e. $\delta_j = 0$ unless $j = 0$ in which case $\delta_0 = 1$. The factor $\frac{\eta}{3}\big(3 + (-1)^j - \delta_{j-1}\big)$ comes from the Simpson's rule that improves the convergence speed towards an integral. Here $e^{ibv_j}\psi(v_j)\frac{\eta}{3}\big(3 + (-1)^j - \delta_{j-1}\big)$ plays the role of the $x(j)$ in 3.6, and we can therefore directly apply the FFT algorithm to the formula 3.8, which gives us the option prices for a whole range of log strikes.

18

3.2 A Fourier cosine series method for option pricing

Another relatively simple method has been introduced in F.Fang and Oosterlee (2008). It offers various advantages with respect to the Carr-Madan method presented earlier. The first one is that this so-called *cos* method does not have restrictions on the strikes over which it computes the options' values and can still compute the prices for various given strikes without recomputing everything. One drawback of the C-M method in option pricing, is that it requires a lot of log strikes to be considered in order to have some accuracy, and as a result, a very low percentage of the option prices is relevant. The *cos* method also avoids the problem of interpolation of the option prices, as the exact strike can always be chosen. Another advantage lies in the fact that where the C-M method only allows for pricing european options, the *cos* method can be used to price several exotic options.

In order to price options assuming the underlying follows a process with analytically-known characteristic function, the idea is to express the risk-neutral density function in terms of the characteristic function, using the inverse Fourier transform, and then to develop the expression in Fourier-cosine series. Once this is done, we can approximate (due to the truncation of the series) the density function in 3.1 (for details about it, see F.Fang and Oosterlee (2008)). The formula of the call price of a European option is

$$C_T(K) \approx Ke^{-rT}\mathrm{Re}\left(\sum_{k=0}^{N-1}{}'\Phi_T\left(\frac{k\pi}{b-a}\right)U_k e^{ik\pi\frac{x-a}{b-a}}\right), \qquad (3.9)$$

where the sign $\sum'$ means that the first term of the sum is divided by 2, and where

$$U_k = \frac{2}{b-a}(\chi_k(0,b) - \psi_k(0,b)),$$

$$\chi_k(c,d) = \left(1 + \left(\frac{k\pi}{b-a}\right)^2\right)^{-1} \left(\cos\left(k\pi\frac{d-a}{b-a}\right)e^d - \cos\left(k\pi\frac{c-a}{b-a}\right)e^c\right.$$

$$\left. + \frac{k\pi}{b-a}\sin\left(k\pi\frac{d-a}{b-a}\right)e^d - \frac{k\pi}{b-a}\sin\left(k\pi\frac{c-a}{b-a}\right)e^c\right),$$

$$\psi_k(c,d) = \frac{b-a}{k\pi}\left(\sin\left(k\pi\frac{d-a}{b-a}\right) - \sin\left(k\pi\frac{c-a}{b-a}\right)\right),$$

unless $k = 0$ in which case $\psi_0(c,d) = d - c$. The parameters a and b are the truncation parameters of the inverse Fourier integral, which is computed over $\mathbb{R}$. Based on the paper of Fang and Oosterlee, we take

$$a = c_1 - L\sqrt{c_2 + \sqrt{c_4}} \text{ and}$$

$$b = c_1 + L\sqrt{c_2 + \sqrt{c_4}},$$

where the c_i are the i-th cumulant and L is a parameter whose value has been chosen to be 10 according to F.Fang and Oosterlee (2008), p. 5. The cumulants for the considered models are given in chapter 6.

Chapter 4

Description of the considered models

As we are now in possession of two pricing methods for European options, we need to choose the models. Among the universe of exponential Lévy processes whose characterstic function is known, we have selected two popular ones : the Variance Gamma and the Merton's jump diffusion models. For these ones, we give the main features, i.e. the stock process or the risk-neutral price dynamics, the characteristic function Φ_T of $\ln S_T$, and we explain on which specific parameters they depend.

First of all, we introduce the parameters common to both models :

- r, the continuous interest rate,

- q, the continuous dividend yield,

- S_0, the stock price at time $t = 0$,

- T the time to expiration of the option in years.

The last parameter not intervening in the price dynamics but in the price of the option is K, the strike.

4.1 The Variance Gamma model

The Variance Gamma model, has been introduced in Carr et al. (1998). It is a Lévy process obtained by estimating a Brownian motion W at random time : given a Brownian motion W with parameters θ and σ, being respectively its drift and standard deviation, the process $\left(X_t(\sigma, \theta, \nu)\right)_{t \geq 0}$ defined by

$$X_t(\sigma, \theta, \nu) = W_{\gamma_t(1,\nu)}(\sigma, \theta)$$

is a VG process. The parameters of the γ process are the mean, 1, and the variance, ν, of the increments per unit of time. Due to the mean of 1, and on average, the new time process will behave like the linear time, which implies that on the long run we do not observe either an acceleration or a deceleration of time. A strictly positive parameter ν ensures that the VG process differs from the Brownian motion. This approach is important as using stochastic time with Lévy processes (instead of a Brownian motion in the present case) is one way to introduce stochastic volatility.

It is worth noting that the VG process can also be expressed as the difference of two independent gamma processes (thus depending on four parameters *a priori*, but of course reduced to three) for simulation purposes. These parameters are known as C, G, and M and given in next subsection.

In addition to Black-Scholes, the VG model allows the skewness of the returns to be different from zero, negative in practice, and explains that the probability of a high loss is higher than that of a high gain. It also takes into account a non-normal kurtosis causing fat tails. However, this model does not explain any leverage effect, which requires a kind of relationship between returns and volatility, for example through the use of stochastic volatility.

We also notice that the stock process to be considered is not the VG process itself but instead the exponential VG process. This is similar to B-S not using the Brownian motion but the geometric Brownian motion for its stock process, which can be seen as an exponential Brownian motion. The stock process will be

$$S_t = S_0 e^{L_t},$$

where $(L_t)_{t \in [0,T]}$ is a VG process plus a drift component.

We finally highlight the fact that the VG process has another interesting feature : a closed-form expression exists for option prices, but that expression is significantly more demanding in terms of computing time than the CM or the *cos* methods.

4.1.1 Parameters of the Variance Gamma model

There are 3 parameters in the VG model :

- $\sigma \in \mathbb{R}_0^+$, the volatility of the Brownian motion on which is built the VG process,

- $\theta \in \mathbb{R}$, the drift of this Brownian motion,

- $\nu \in \mathbb{R}_0^+$, the volatility of the γ time-changing process.

Alternatively, the three other parameters C, G and M are related to σ, θ and ν via the following formulas (Carr et al., 2003, p. 350) :

$$C = \frac{1}{\nu},$$

$$G = \left(\sqrt{\frac{\theta^2 \nu^2}{4} + \frac{2\sigma^2 \nu}{2}} - \frac{\theta \nu}{2} \right)^{-1},$$

$$M = \left(\sqrt{\frac{\theta^2 \nu^2}{4} + \frac{2\sigma^2 \nu}{2}} + \frac{\theta \nu}{2} \right)^{-1}.$$

4.1.2   Characteristic function of the VG model

In the case of an exponential VG process with parameters σ, θ and ν, the characteristic function of $\ln S_t$ is expressed as

$$\Phi_t(u) = \left(1 - i\theta\nu u + \frac{\sigma^2 \nu u^2}{2} \right)^{-t/\nu}.$$

To have a risk-neutral process, we need to rectify it by a drift ω. Consequently, the stock price at time t is given by :

$$S_t = S_0 \exp\big((r + \omega)t + X_t(\sigma, \theta, \nu)\big),$$

with

$$\omega = \frac{1}{\nu} \ln(1 - \theta\nu - \frac{1}{2}\sigma^2 \nu).$$

As we consider the risk-neutral dynamics, the characteristic function to be considered is then :

$$\Phi_t(u) = \exp\big(iu[\ln(S_0) + (r + \omega)t]\big) \left(1 - i\theta\nu u + \frac{\sigma^2 \nu u^2}{2} \right)^{-t/\nu}. \qquad (4.1)$$

4.2   The Merton's jump diffusion model

This model has been proposed for the first time in Merton (1976). As the VG model, it is a pure exponential Lévy process and its stock process can be expressed

24

as

$$S_t = S_0 e^{L_t}$$

with $(L_t)_{t \in [0,T]}$ being a Lévy process. In the case of the Merton's model, the Lévy process L_t is a Brownian motion with drift, just as in Black-Scholes, together with a compound Poisson process, the jump component. The price dynamics is given by

$$\frac{\mathrm{d}S_t}{S_t} = (r - q - \frac{\sigma^2}{2} - \lambda \mu_J)\mathrm{d}t + \sigma \mathrm{d}W_t + J_t \mathrm{d}N_t.$$

There are three independent processes : W a standard Brownian motion with volatilty σ, $N = \{N_t; t \geq 0\}$ a Poisson process with parameter $\lambda > 0$, and $J_t := y_t - 1$, a lognormal random variable, identically distributed and independent over time. μ_J is the unconditional mean of J_t and σ_J is the standard deviation of $\ln(1 + J_t)$. As a consequence, $\ln(y_t)$ is a normal variable and we will consider in the following the parameters μ and δ^2 such that $\ln(y_t) \sim \mathcal{N}(\mu, \delta^2)$ instead of the parameters μ_J and σ_J^2. Of course both parametrisations are equivalent and are linked by the following formulas:

$$\mu_J = e^{\mu + \frac{1}{2}\delta^2} - 1,$$

$$\sigma_J^2 = e^{2\mu + \delta^2}(e^{\delta^2} - 1).$$

4.2.1 Parameters of the Merton's model

There are 4 parameters in the Merton's model :

- $\sigma \in \mathbb{R}_0^+$, the volatility of the Brownian motion component,

- $\lambda \in \mathbb{R}^+$, the parameter of the Poisson process N_t,

- $\mu \in \mathbb{R}$, the mean of $\ln(y_t)$,

- $\delta \in \mathbb{R}_0^+$, the standard deviation of $\ln(y_t)$.

4.2.2   Characteristic function of the Merton's model

The characteristic function of the risk-neutral Merton process is given by Matsuda (2004), p. 10 :

$$\Phi_t(u) = \exp\left[t\left(ibu - \frac{\sigma^2 u^2}{2} + \lambda\left(\exp\left(i\mu u - \frac{\delta^2 u^2}{2}\right) - 1\right)\right)\right], \qquad (4.2)$$

where $b = r - q - \frac{\sigma^2}{2} - \lambda\mu_J$.

Chapter 5

Calibration methodology

Model calibration to real data consists in determining the parameters of the models that best fit relevant data. One way to tackle this problem, in the case of a stock index for example, is to estimate the desired parameters based on the past data of the index. For the Black-Sholes model, it would be taking as a proxy for the instantaneous volatility the historical one. Even though this approach seems to make sense, several problems arise. How far in the past should we look back? Aren't recent moves more relevant than older ones? In fact, as commented in Cont and Tankov (2004), chapter 13, in incomplete markets, historical data leads to different prices for a European option. For that reason, calibration is not performed based on past data but instead using current "relevant" data, in practice option prices. In the B-S model, the input is an option price, given its underlying price, its strike, the interest rate and the time to maturity. The *inverse* B-S formula then returns the implied volatility, the volatility the option should have in order to have the given price according to the B-S model, because it is a one parameter model. It is important to notice that calibration is not as straighforward as it might seem due to the fact that based on option prices, one needs to find the parameters inherent to the model and not the contrary. That is why it is called an *inverse* problem.

In general, however, models rely on more than one parameter and the calibration is realised by considering a set of call options $\mathcal{C}$ having different strikes for a given maturity and by choosing the model's parameters that best approximate the prices. In practice, models with a few parameters do not return the exact price of each option in $\mathcal{C}$, and that is why the aim of calibration is to minimise an "error" function depending on the parameters that expresses "how far" the prices predicted by the model are from the real prices. Another reason to be cautious with the calibration is that even if the pricing formula is continuous in each of its parameters, the contrary does not hold in general (making the calibration problem an *ill-posed* problem in the sense of Hadamard), i.e., a very small change in the initial conditions, or option prices in our case, may lead to a big jump in the estimated parameters, and it is well known that noise exists in the options' market. As a more liquid option will ensure less noise, the set $\mathcal{C}$ will be a set of liquid options, European options most of the time. In some exceptional cases, exotic options may also be used for calibration.

Even though we assume the market prices to be the correct ones, calibration is useful as it allows to price other assets, through the use of Monte Carlo simulations for example, such as unpriced exotic options. The choice of an adequate model is relevant because in order to price specific products, we need to use a model taking into account the specific risks of the product.

5.1   The calibration formula

The choice of a good calibration formula, or the choice of the error function to be minimised, is primordial as different plausible formulas may possibly lead to very different parameters. We discuss in this section such a choice. Let $C_i \in \mathcal{C}$, $i \in \{1, 2, ..., N\}$, $N \in \mathbb{N}$, denote the market prices of the considered European options for a given maturity T and let K_i denote their corresponding strikes. The model's parameters will be called θ. Calibration allows to choose the optimal parameters θ^* by minimising a weighted least square function :

$$\theta^* = \arg \inf_{\theta \in \Omega} \sum_{i=1}^{N} w_i \big(C_i - C(K_i; \theta)\big)^2, \tag{5.1}$$

where $C(K_i; \theta)$ is the option price for the considered pricing model with parameters θ and strike K_i, and Ω denotes the bounded set of the possible values for the parameters. Apart from determining the set $\mathcal{C}$, which will be done in the next chapter, the w_i's need to be chosen, and this leads to two different approaches. The first one is proposed in Tankov (2009), p. 90 : in order not to overweight options with higher prices, we will first set

$$w_i = \frac{1}{C_i^2}. \tag{5.2}$$

The second one, proposed in Cont and Tankov (2004), chapter 13, is :

$$w_i = \text{vega}_{\text{BS}}^2(\hat{\sigma}_i), \tag{5.3}$$

where $\hat{\sigma}_i$ denotes the BS implied volatility for the option i. This formula is a proxy to the one minimising the sum of the absolute differences between the B-S implied volatilities of the market and the ones of the model. Using the explicit function would be very slow as it implies computing a new implied volatility at each step of the minimisation. The weights in 5.3 offer the advantage of approximating the implied volatility, which is often relevant in practice.

Chapter 6

Computational precisions

All the computations have been performed using the Matlab version 7.9.0.529 (R2009b) 64-bit on a Mac 2.5 GHz Intel Core i5 with 4 GB 1333 MHz DDR3 of memory. With the CM approach, we compute the options' prices assuming that the price of the underlying at time $t = 0$ is 1 and then, we multiply the result by S_0. The reasons behind that choice are clear : simplicity and computational efficiency. We are allowed to do this because the considered processes are scale invariant and one of the characterisations of that property is that the price of a European call is homogenous of degree one with respect to S_0 and K :

$$aC_T(S_0, K, \theta) = C_T(aS_0, aK, \theta) \quad \forall a > 0,$$

where θ indiquates the same parameters. To see that a process is scale invariant, it is enough to observe that the characteristic function of the log prices is independent of S_0, another characterisation of the scale invariance property. For more details, see Alexander and Nogueira (2005) or Fengler (2012), chapter 10.

6.1 Interpolation for the Carr-Madan method

Using the CM method, we usually want to estimate an option price at a given strike which does not belong to the strike grid resulting from 3.8. It is not a problem for the *cos* method but the CM method only allows for equidistant log strikes. That is why a given strike K_i is not likely to coincide with one strike considered by the CM method. Thus, interpolation is needed in order to fill the gaps and we consider two different methods. The first one is straightforward and still often used in practice, even if it lacks some precision : the linear interpolation.

The second one is a bit more sophisticated and interpolation is performed via the use of natural cubic splines. To do so, we utilise an efficient algorithm given in Fengler (2012), chapter 6, described in the following. Let us suppose that we are given n points $(u_1, g_1), (u_2, g_2), ...(u_n, g_n)$ on the plan $\mathbb{R}^2$ that we want to interpolate via the use of a natural cubic spline, i.e. a function g which is a polynom of degree 3 on each segment $[u_i, u_{i+1}]$, g, g' and g'' being continuous and the function being a polynom of order 1 outside $[u_1, u_n]$. Let $\gamma_i = g''(u_i)$, then the only natural cubic spline interpolating the n points is determined by the vectors $\mathbf{g} = (g_1, ..., g_n)^T$ and γ, the first element of the vector γ being γ_2 and its last one being γ_{n-1}, as $\gamma_1 = \gamma_n = 0$. The vector γ is given by :

$$\gamma = R^{-1} Q^T \mathbf{g}.$$

We now describe the two matrices Q and R. Let $h_i = u_{i+1} - u_i \ \forall i = 1, ..., n - 1$. The $n \times (n - 2)$ matrix Q is defined by

$$q_{j-1,j} = h_{j-1}^{-1}, \quad q_{j,j} = -h_{j-1}^{-1} - h_j^{-1}, \quad q_{j+1,j} = h_j^{-1},$$

for $j \in 2, ..., n - 1$, and $q_{i,j} = 0$ for $|i - j| > 1$. Similarly to the γ vector, the top left element of Q is $q_{1,2}$.

The $(n - 2) \times (n - 2)$ matrix R is defined by

$$r_{i,i} = \frac{1}{3}(h_{i-1} + h_i), \quad i \in [2, ..., n - 1] \ \text{ and}$$

$$r_{i,i+1} = r_{i+1,i} = \frac{1}{6} h_i, \quad i \in [2, ..., n - 2].$$

Having computed those values we can define the cubic spline g for $u \in [u_i, u_{i+1}]$ as follows :

$$g(u) = \frac{\gamma_{i+1}}{6h_i}(u - u_i)^3 + \frac{\gamma_i}{6h_i}(u_{i+1} - u)^3$$
$$+ \left(\frac{g_{i+1}}{h_i} - \frac{h_i}{6}\gamma_{i+1}\right)(u - u_i) + \left(\frac{g_i}{h_i} - \frac{h_i}{6}\gamma_i\right)(u_{i+1} - u).$$

On the following graph we can see an illustration of a cubic spline interpolation.

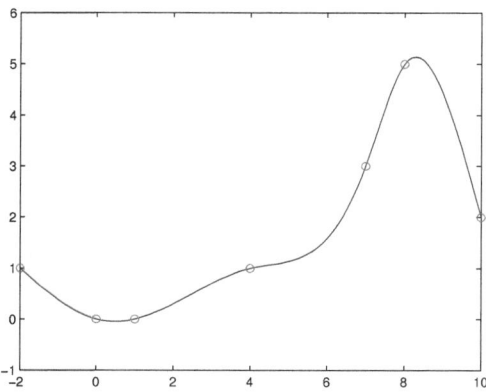

Figure 6.1: Cubic spline interpolation

In practice, we use the output of the CM formula for strikes $K \in [0.6S_0, 1.5S_0]$, approximately 3.66% of the total output, to interpolate with natural cubic splines.

6.2 Remarks about the *cos* method

To be fair with respect to the CM method which computes a lot of strikes at the same time, we computed the value of several options by computing the part independent from the strike only once, using the *cos* method.

6.2.1 Cumulants of the VG model

In the case of the VG model, the cumulants are given as in the following :

$$c_1 = (r + \theta)T,$$
$$c_2 = (\sigma^2 + \nu\theta^2)T,$$
$$c_4 = 3(\sigma^4\nu + 2\theta^4\nu^3 + 4\sigma^2\theta^2\nu^2)T.$$

6.2.2 Cumulants of the Merton's model

In the case of a Merton's model, the cumulants are given as in the following:

$$c_1 = (r - \frac{\sigma^2}{2} - \lambda k + \lambda\mu)T,$$
$$c_2 = (\sigma^2 + \lambda\delta^2 + \lambda\mu^2)T,$$
$$c_4 = \lambda(3\delta^4 + 6\mu^2\delta^2 + \mu^4)T.$$

Chapter 7

Description of the data

All the data about options used for calibration purposes can be found on http://wrds-web.wharton.upenn.edu/wrds/, and were selected using the Option Metrics database. The data used are the S&P500 European call prices on December 21, 2012. We consider time horizons of six, nine and twelve months from that day. For the risk-free rate, we take the continuous interest rate of the zero coupon bond for the corresponding period. All the considered data can be found in the Appendix A.

To have a reliable dataset, we apply some filters. This is a common practice although some skip this step. Even though a part of the data is lost, the benefits from using less numerous but more relevant data make it worth doing it.

Here is a list of the considered filters.

1. We only consider options whose open interest is above 1.000 during the considered day for liquidity purposes.

2. We drop options that are too far from the money, i.e. we only keep options checking $0,8 < \frac{K}{S_0} < 1,2$.

3. Also for liquidity purposes, we drop the options with a bid-ask spread too

high for their prices, i.e. options verifying the following property :

$$\frac{\text{Ask Price} - \text{Bid Price}}{\text{Mid Price}} > 0, 2.$$

4. We eventually drop the options for which no implied volatility is given.

We do not exclude other relevant filters for another set of options, but no more filters were necessary for our dataset. For a more exhaustive list of filters see Constantinides et al. (2011), Appendix A.2.

Chapter 8

Empirical results

We finally compare the different techniques described so far, i.e. the Carr-Madan approach and the *cos* one. Firstly, we determine the parameters of those methods (which are independent of the processes) that offer both acceptable accuracy and computational time. To do so, we simultaneously price three options twice, one ITM, one ATM and one OTM for some fixed parameters and two different times to maturity. We compare the results obtained for the CM approach using both the linear and cubic spline interpolations to the results of the *cos* approach. We finally discuss the calibration of all those different methods using different weights in our calibration : $w_i = \frac{1}{C_i^2}$ and $w_i = \text{vega}_{\text{BS}}^2(\hat{\sigma}_i)$.

8.1 Choice of the parameters for the Carr-Madan method

The parameters related to the CM method we need to determine are N, where N is a power of 2, η and α. The truncation parameter a is equal to $N\eta$ and we assume it constant and equal to 1024, so that the parameter η does not need to be determined. We are left with the parameter α that we set equal to $\alpha = 1.5$ as recommended in Carr and Madan (1999), p. 70. Taking a higher N while keeping

η constant does not make the computation much more accurate and we observed that a value of $\alpha = 1.5$ made the integral converge faster than using other values in the computation we made. We later show that those choices are confirmed by the *cos* method that converges towards the same values as the ones obtained with the CM method and the chosen parameters up to a very small error. We now focus on finding a good N such that the result is close enough to the value of the whole series. The six options we consider have the following characteristics : $S_0 = 100$, $r = 0.03$, $q = 0$, $T = 0.2$ and $T = 1$ with $K = 80$, 100 and 120. For the VG model we use the following parameters:

$$\sigma = 0.2, \ \theta = -0.1, \ \text{and} \ \nu = 0.2,$$

and for the Merton's one we use these ones :

$$\sigma = 0.2, \ \lambda = 0.3, \ , \mu = -0.2, \ \delta = 0.5.$$

The same parameters are used to price options using the *cos* method. On figure 8.1, we see the output of the CM method for a VG process with $T = 1$ for an extended range of strikes, and on figure 8.2, we observe the similar curve for a Merton process with $T = 0.2$. The output of the *cos* method would look exactly the same at the current scale with the same parameters.

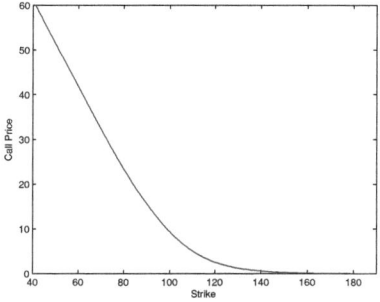

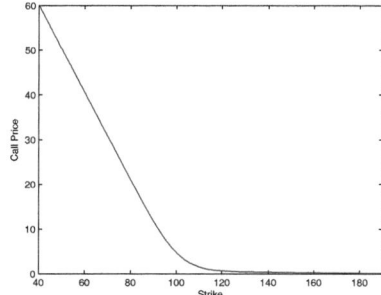

Figure 8.1: Price curve for the VG model with $T = 1$

Figure 8.2: Price curve for the Merton's model with $T = 0.2$

The results for the CM method can be found in the Appendix B (where N denotes the exponent of 2 i.e. $N = 12$ means that $N = 2^{12}$) and we point out several relevant observations. First of all, the accuracy is good enough for $N = 4096$ as the error between the approximation and the true value is less than 10^{-6} (with $S_0 = 100$, meaning $S_0 e{-}08$) either for a short ($T = 0.2$) or a long ($T = 1$) maturity, independently of the stock process. We also notice that the convergence is slower for a shorter maturity than for a longer one. Finally, the cubic spline interpolation has more or less the same rate of convergence as the linear interpolation which could be expected as the spaces between the strikes are small. A value of $N = 8192$ could be used in case an accuracy of less than $S_0 e - 11$ is required. We should not choose N greater than 8192 because the convergence is then very slow. The CPU times to compute the options' prices are almost equal for the CM method with linear interpolation and the *cos* one but are much higher while using the CM method with cubic spline interpolation.

8.2   Choice of the parameters for the *cos* method

The same methodology is applied to find the right truncation parameter N to use in the formula 3.9. The considered options will have the same characteristics as the ones used in last section : $S_0 = 100$, $r = 0.03$, $q = 0$, $T = 0.2$ and $T = 1$ with $K = 80$, 100 and 120.

The VG parameters are

$$\sigma = 0.2, \ \theta = -0.1, \ \text{and} \ \nu = 0.2,$$

and the Merton ones are

$$\sigma = 0.2, \ \lambda = 0.3, \ ,\mu = -0.2, \ \delta = 0.5.$$

The results of the calibration are given in the Appendix C. According to the results, we notice that a parameter $N = 160$ is fine for options having a maturity of one year as it allows for an accuracy of $S_0 e - 09$ at least for both processes. Nonetheless, and as convergence is slower for options having shorter maturities, we recommend using a higher N for options checking $T \leq 0.2$ for the VG process in particular. As we calibrate the models to market options whose time to maturity is at least half a year, we use the value $N = 160$.

8.3   Linear interpolation vs cubic spline interpolation

Does a cubic spline interpolation really add accuracy over a linear interpolation and should we use a cubic spline interpolation? The advantage of using cubic splines is that it respects the positive convexity of the price curve observed on figures   8.1 or   8.2. Therefore, it requires the vector γ (see chapter 6) to be positive in all its components, or it would offer theoretical arbitrage opportunities. We now consider the VG model with exactly the same parameters as before. The figure 8.3 represents the differences between the cubic spline interpolation and the linear one around $K = 80$, with the blue crosses being the points that are not interpolated.

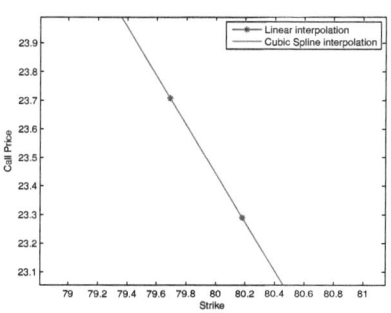

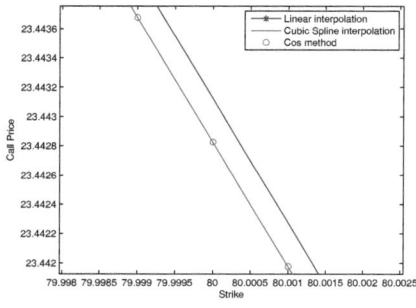

Figure 8.3: Linear interpolation vs CS interpolation around $K = 80$

Figure 8.4: Linear interpolation vs CS interpolation around $K = 80$ zoomed

Figure 8.3 shows that both lines are pretty similar if not equal due to the relatively fine strike grid. However, zooming more (see figure 8.4) highlights that those lines are different, and that the cubic spline is below the straight line, fact that is justified by the positive convexity (or the positive γ_i's) of the curve.

Figures 8.5 and 8.6 show that the same phenomenon is also valid for a strike of $K = 120$. Those graphs confirm that there is a loss of accuracy while using linear interpolation, that is of the order of $S_0 e - 05$ for a strike somewhere in the middle of two strikes computed by the CM method. More important, the results obtained using the cubic spline interpolation almost exactly coincide with the ones given by the *cos* method, which emphasises that the *cos* method is slightly more accurate than the CM one with linear interpolation, but does not offer any practical accuracy advantage over the CM method with cubic spline interpolation. We also notice that using the CM method with linear interpolation and $N = 4096$, the precision is of about $S_0 e - 05$ compared to the more accurate CM method with cubic spline interplation or to the *cos* method. This last point implies that there is no need to use $N \geq 8192$ if linear interpolation is chosen, as considering increasing N, the options' prices will converge towards biased values. A value of $N = 2048$ may be too small as the bias of the options' prices due to the linear interpolation, more or less $S_0 e - 05$, is smaller than the error due to the N which is of about $2.68e - 05$.

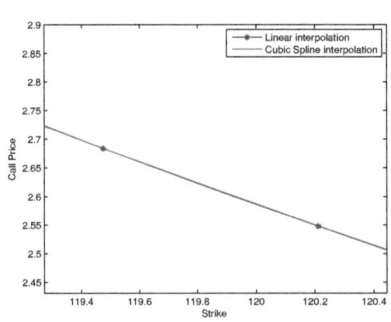

Figure 8.5: Linear interpolation vs CS interpolation around $K = 120$

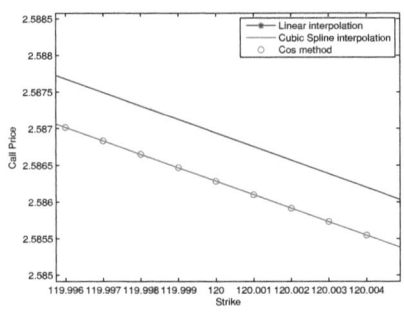

Figure 8.6: Linear interpolation vs CS interpolation around $K = 120$ zoomed

Let us mention at this point that the same phenomenon holds for the Merton's model.

8.4 Calibration of the models

We now calibrate the VG and the Merton's models to the empirical data described earlier. For efficiency purposes, we do not perform our code for a huge number of parameters and choose the ones returning the lowest error with respect to real prices, but instead use an optimisation under constraints algorithm. As a matter of fact, unconstrained optimisation returns a price curve further from reality than the constrained one but also sometimes proposes parameters that do not make any sense.

Two problems arise while adopting this method. The first one is that the parameters θ^* that optimise 5.1 are very sensitive to the initial parameters and to the constraints we choose (as we cannot simply decide for example that $\sigma \in \mathbb{R}^+$ but instead $\sigma \in [\epsilon, B]$, $B < +\infty$, implying that ϵ and B need to be determined). In other words, the solution is dependent on the input, which is often the case when the function we want to optimise allows for local minima.

The second problem is that reasonably different parameters may lead to two price curves that almost equally approximate the market prices.

After testing several times for good inputs, we decided to keep the following limits for the VG process: $\sigma \in [0.001, 10]$, $\theta \in [-10, 10]$, $\nu \in [0.001, 10]$. For the Merton one we chose : $\sigma \in [0.001, 10]$, $\lambda \in [0.001, 10]$, $\mu \in [-10, 10]$ and $\delta \in [0.001, 10]$.

In the Appendices D and E, we report the parameters obtained for the three methods, both processes and both weights. We want to point out that the same 25 starting points for the minimisation have been used for the CM with linear interpolation and the *cos* methods, but we only chose 5 of them for the CM method with cubic splines interpolation as it was already sufficiently slow.

The observations can be summed up in the following.

1. In terms of rapidity, the CM method with cubic spline interpolation is much slower than the two other ones. Based on calibration time, it is hard to say whether the CM with linear interpolation method is faster than the *cos* one

as the time elapsed are pretty similar, once in favour of the first method, once in favour of the other one.

2. Using several starting points for the optimisation is relevant. Sometimes it will return results with 80% of good candidates (which means parameters that return an error very close to the smallest one observed), but sometimes only 25% will be good candidates.

3. We hardly observe a difference in the estimated parameters depending on the pricing method. They usually coincide up to the second or third digit. Some exceptions arise however and, in the case of the CM method with cubic spline interpolation, they are due to the fact that we have only used five starting points, resulting in less accurate estimated parameters. A more relevant exception is observed for the Merton model using the vega weights, that may be due to the fact that the *cos* method is slightly more accurate.

4. Both models accurately approximate the market prices of the options, and we cannot clearly decide that one should be used instead of the other one based on European options.

5. Using $\frac{1}{\text{vega}^2}$ and $\frac{1}{P_i^2}$ as weights returns noticeably different parameters, but all those parameters give a good approximation of the market prices. Nevertheless, most of the time, the $\frac{1}{P_i^2}$ weights offer a better approximation of the options' prices, as the $\frac{1}{\text{vega}^2}$ weights offer a better one of the implied volatility curve.

6. The implied volatility curves given by the models fit quite well the implied volatility of the options in our dataset. We should however be careful about the extrapolation of the implied volatility curves outside the strikes of the dataset, as they can have very different shapes depending on the chosen parameters as noticed on figure 8.12. Consequently, both models, given well

chosen parameters, are able to explain the volatility smile observed on the market or the kurtosis.

7. While using the cubic spline interpolation, we have never observed any problem relating to the gamma's that were always positive.

8. The negative skewness is explained in the VG model by the negative θ, and in the Merton's model by the negative μ_J, the expectation of the jump, observed in the calibration.

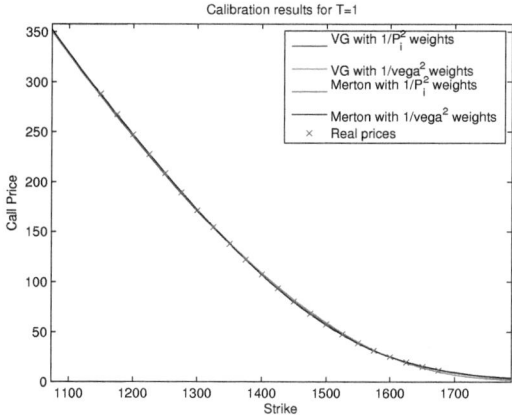

Figure 8.7: Calibration results for T=1

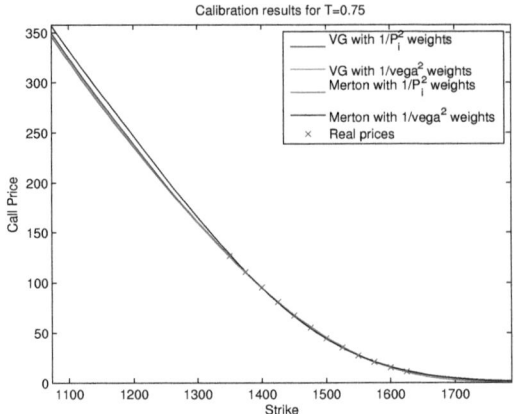

Figure 8.8: Calibration results for T=0.75

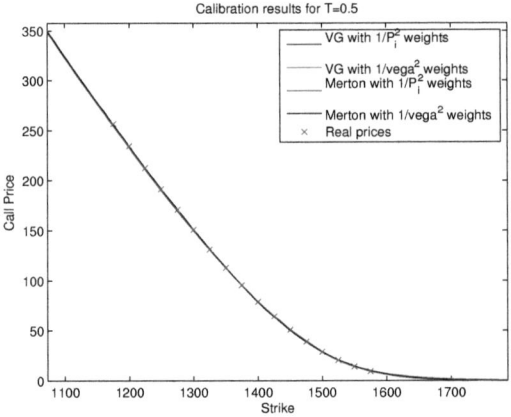

Figure 8.9: Calibration results for T=0.5

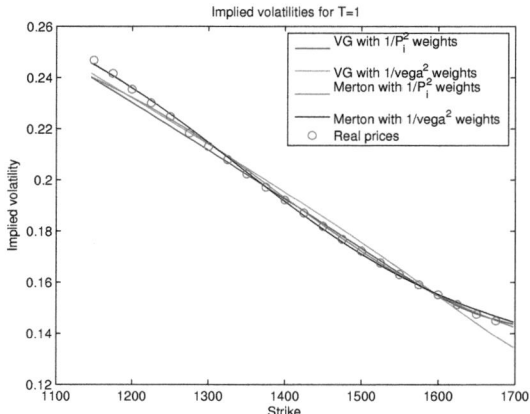

Figure 8.10: Calibration results for T=1

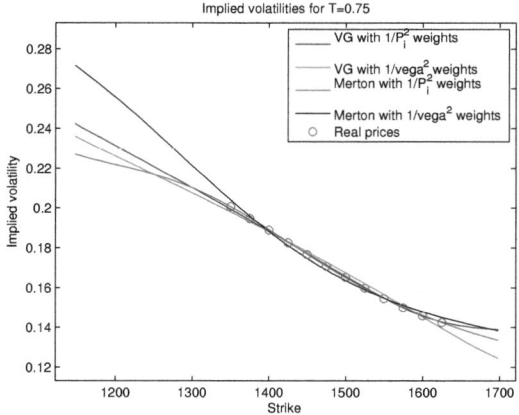

Figure 8.11: Calibration results for T=0.75

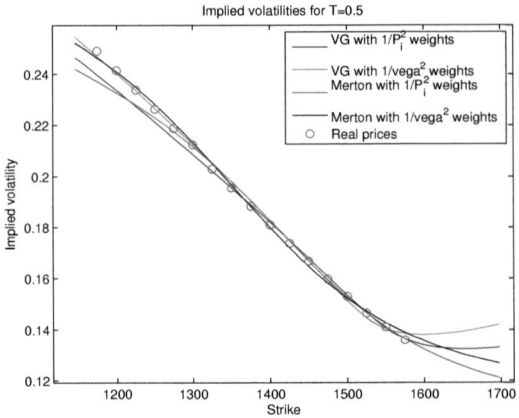

Figure 8.12: Calibration results for T=0.5

Chapter 9

Conclusion

After comparing two or three pricing methods, depending on how we want to count them, we noticed that the *cos* one has an edge over the other two, both in precision and in computational time. However, for calibration purposes, the gain in precision of the *cos* method, over the CM one with linear interpolation, is barely useful. Both the VG and Merton's models offer good calibration results, even if Merton's one looks a bit more accurate, mainly due to the fact that this is a 4-parameter model and that the VG one is a 3-parameter model. The use of weights $w_i = \frac{1}{\text{vega}^2}$, instead of the more common $w_i = \frac{1}{P_i^2}$, is relevant to better estimate the i.v. without computing an i.v. at each step, which would require a lot of time. Approximating the i.v. makes sense, since options are often priced in terms of implied volatility.

In future, one could consider more advanced pricing methods as the fractional Fourier transform or Lewis' approach (see Schmelzle (2010) for these methods), or even the use of caching techniques (see Kilin (2011)) in order to decrease the computational time and improve precision. Also, to make it more realistic - regarding volatility clustering or the leverage effect - one could apply the presented techniques to time-changed Lévy processes or processes involving stochastic volatility. A final point, often not discussed in literature, and that one would certainly require a more in-depth study, concerns the data used for calibration purposes. A pricing method could be perfect, using it is worthless in case the dataset assumed to be "correct" is not suitable.

Bibliography

C. Alexander and L. Nogueira. Optimal Hedging and Scale Invariance : A Taxonomy of Option Pricing Models. *Discussion paper, University of Reading*, 2005.

G. Bakshi and D. Madan. Spanning and derivative-security valuation. *Journal of Financial Economics*, 55:205–238, 2000.

F. Black and M. Scholes. The Pricing of Options and Corporate Liabilities. *The Journal of Political Economy*, 81:637–654, 1973.

P. Carr and D. B. Madan. Option valuation using the Fast Fourier transform. *Journal of Computational Finance*, 2:61–73, Fall 1999.

P. Carr and L. Wu. Time-changed Lévy processes and option pricing. *Journal of Financial Economics*, 71:113–141, 2004.

P. Carr, E. Chang, and D. Madan. The Variance Gamma Process and Option Pricing. *European Finance Review*, 2:79–105, 1998.

P. Carr, H. Geman, D. B. Madan, and M. Yor. Stochastic Volatility for Lévy processes. *Mathematical Finance*, 3:345–382, July 2003.

G. H. L. Cheang and C. Chiarella. A modern view on Merton's Jump-Diffusion Model. *Quantitative Finance Research Centre*, Research paper 287, 2011.

Z. Chen, X. Lin, and L. Feng. Simulating Lévy processes from their characteristic functions and financial applications. *ACM Transactions on Modeling and Computer Simulation*, 22, 2012.

G. Constantinides, J. Jackwerth, and A. Savov. The Puzzle of Index Option Returns. *Working paper, University of Chicago, University of Konstanz, and New York University*, 2011.

R. Cont and P. Tankov. *Financial modelling with Jump processes.* Chapman & Hall / CRC Financial Mathematics Series, 2004.

J. C. Cox, J. E. Ingersoll, and S. A. Ross. A Theory of the Term Structure of Interest Rates. *Econometrica*, 53:385–408, 1985.

A. Czerny. Introduction to Fast Fourier transform. *The Journal of Derivatives*, Fall 2004.

M. Fengler. Advanced Topics in Pricing and Hedging of Equity Derivatives. *PhD Course, St. Gallen University*, 2012.

F.Fang and C. W. Oosterlee. A novel pricing method for European options based on Fourier-cosine series expansions. *SIAM J. Sci. Comput.*, 31:826–848, 2008.

S. L. Heston. A closed-form solution for options with stochastic volatility with applications to bonds and currency options. *The Review of Financial Studies*, 6(2):327–343, 1993.

A. Itkin. Pricing options with VG model using FFT. *arxiv.org/pdf/physics/0503137*, 2010.

F. Kilin. Accelerating the calibration of stochastic volatility models. *The Journal of Derivatives*, pages 7–16, Spring 2011.

K. Matsuda. Introduction to Merton Jump Diffusion Model. *The Graduate Center, The City University of New York*, 2004.

R. C. Merton. Option pricing when underlying stock returns are discontinuous. *Journal of Financial Economics*, 3(1):125–144, 1976.

M. Schmelzle. Option Pricing Formulae using Fourier Transform : Theory and Application. 2010.

W. Schoutens, E. Simons, and J.Titsaert. A Perfect Calibration ! Now what ? *Wilmott Magazine*, March 2005.

S. E. Shreve. *Stochastic Calculus for Finance II, Continuous-Time Models.* Springer, 2004.

P Tankov. Calibration de Modèles et Couverture de Produits Dérivés. *Polycopié du cours des masters Modélisation Aléatoire de l'Université Paris-Diderot (Paris 7) et Probabilités et Finances de l'Université Pierre et Marie Curie (Paris 6),* 2009.

P. Tankov. Financial modelling with Lévy processes. *Notes of lectures given by the author at the Institute of Mathematics of the Polish Academy of Sciences,* October 2010.

P. Tankov and E. Voltchkova. Jump-diffusion models : a practitioner's guide. *Banques et Marchés*, 99, 1999.

N. F. Kouemo Tchamga. Fourier Transform Methods in Option Pricing. 2009.

Appendix A : Option data

Table 1: Options quotes on 21 December, 2012

Maturity	22/06/13		21/09/13		21/12/13	
Strike	Bid	Ask	Bid	Ask	Bid	Ask
1150					285,6	290,3
1175	254,8	259,3			265	270
1200	232,6	237			244,8	249,5
1225	210,8	215,2			225,3	229,9
1250	189,6	193,9			206,2	210,9
1275	168,9	173,2			187,8	191,3
1300	149,3	153,5			170	173,4
1325	130,2	132,7			152,7	156,5
1350	111,8	114,4	125,7	128,6	136,2	139,7
1375	94,5	97	109,3	112,3	120,5	124,1
1400	78,3	80,8	94,3	96,7	105,9	109,1
1425	63,4	65,9	79,6	82,3	91,8	95,2
1450	50,1	52,4	66,3	69	78,8	82,1
1475	38,2	40,6	54,3	56,8	66,7	70,1
1500	28,2	30,3	43,7	46	56,3	58,9
1525	20,1	21,6	34,2	36,6	46,5	49
1550	13,6	15,2	26,2	28,4	38,1	40
1575	8,8	10,2	19,9	21,5	30,6	32,6
1600			14,3	16,4	24,2	26,1
1625			10,3	12,2	18,5	20,8
1650					13,8	16,3
1675					10,5	12,7

The dividend yield is 2.346257%. The annualised risk-free rates are 0.305103%, 0.318356% and 0.33004% for maturities of six months, nine months and one year respectively.

Appendix B : Errors and CPU times for the Carr-Madan method

Table 2: For a VG process with T=1 and linear interpolation

N	K=80	K=100	K=120	CPU time (in seconds)
10	-2.8136e-01	-2.7885e-01	-2.7634e-01	7.8751e-04
11	-2.6854e-03	-2.6845e-03	-2.6835e-03	0.0012
12	-2.1705e-07	-2.1706e-07	-2.1707e-07	0.0024
13	3.1768e-11	1.9190e-11	4.9769e-12	0.0046
14	3.1658e-11	1.9151e-11	5.0502e-12	0.0072

Table 3: For a VG process with T=0.2 and linear interpolation

N	K=80	K=100	K=120	CPU time (in seconds)
10	-2.8112e-01	-2.7855e-01	-2.7598e-01	7.0233e-04
11	-2.6853e-03	-2.6843e-03	-2.6833e-03	0.0013
12	-2.1791e-07	-2.1761e-07	-2.1752e-07	0.0024
13	-4.2169e-10	-2.3844e-10	-2.1869e-10	0.0039
14	-2.1907e-10	-9.3414e-11	-1.0832e-10	0.0067

Table 4: For a VG process with T=1 and cubic spline interpolation

N	K=80	K=100	K=120	CPU time (in seconds)
10	-2.8136e-01	-2.7885e-01	-2.7634e-01	0.0148
11	-2.6854e-03	-2.6845e-03	-2.6835e-03	0.0169
12	-2.1707e-07	-2.1707e-07	-2.1707e-07	0.0184
13	4.8992e-12	4.7784e-12	6.7097e-12	0.0216
14	4.7748e-12	4.7393e-12	6.7866e-12	0.0246

Table 5: For a VG process with T=0.2 and cubic spline interpolation

N	K=80	K=100	K=120	CPU time (in seconds)
10	-2.8112e-01	-2.7855e-01	-2.7598e-01	0.0157
11	-2.6853e-03	-2.6843e-03	-2.6833e-03	0.0166
12	-2.1788e-07	-2.1765e-07	-2.1752e-07	0.0180
13	-3.9857e-10	-2.7621e-10	-2.1981e-10	0.0216
14	-1.9599e-10	-1.3118e-10	-1.0943e-10	0.0241

Table 6: For a Merton process with T=1 and linear interpolation

N	K=80	K=100	K=120	CPU time (in seconds)
10	-2.8629e-01	-2.8073e-01	-2.7719e-01	7.3974e-04
11	-2.6853e-03	-2.6846e-03	-2.6835e-03	0.0014
12	-2.1709e-07	-2.1705e-07	-2.1707e-07	0.0029
13	-8.2174e-12	3.0132e-11	1.3999e-11	0.0054
14	-8.2885e-12	3.0102e-11	1.4072e-11	0.0081

Table 7: For a Merton process with T=0.2 and linear interpolation

N	K=80	K=100	K=120	CPU time (in seconds)
10	-2.8117e-01	-2.7856e-01	-2.7598e-01	7.5413e-04
11	-2.6853e-03	-2.6843e-03	-2.6833e-03	0.0015
12	-2.1699e-07	-2.1699e-07	-2.1707e-07	0.0028
13	8.8797e-11	8.8722e-11	9.2693e-12	0.0052
14	8.8651e-11	8.8678e-11	9.2841e-12	0.0081

Table 8: For a Merton process with T=1 and cubic spline interpolation

N	K=80	K=100	K=120	CPU time (in seconds)
10	-2.8629e-01	-2.8073e-01	-2.7719e-01	0.0149
11	-2.6858e-03	-2.6846e-03	-2.6835e-03	0.0178
12	-2.1708e-07	-2.1707e-07	-2.1708e-07	0.0196
13	2.6397e-12	7.5104e-12	3.4905e-13	0.0213
14	2.5473e-12	7.4802e-12	4.2633e-13	0.0251

Table 9: For a Merton process with T=0.2 and cubic spline interpolation

N	K=80	K=100	K=120	CPU time (in seconds)
10	-2.8117e-01	-2.7856e-01	-2.7598e-01	0.0160
11	-2.6853e-03	-2.6843e-03	-2.6833e-03	0.0164
12	-2.1707e-07	-2.1707e-07	-2.1708e-07	0.0183
13	5.1479e-12	1.1049e-11	9.0705e-14	0.0211
14	4.9738e-12	1.1005e-11	1.0536e-13	0.0260

Appendix C : Errors and computational times for the *cos* method

Table 10: For a VG process with T=1

N	K=80	K=100	K=120	CPU time (in seconds)
20	2.8767e-01	-2.6426e-01	-7.2601e-02	3.7704e-04
50	-2.9809e-04	-3.7313e-04	-5.0501e-05	8.8387e-04
100	2.2169e-07	2.9822e-07	-5.0517e-08	0.0017
160	6.4136e-10	6.4521e-10	-3.8075e-11	0.0027
200	-3.9627e-11	-2.9457e-11	-7.5743e-11	0.0034
1000	0	0	0	0.0165

Table 11: For for a VG process with T=0.2

N	K=80	K=100	K=120	CPU time (in seconds)
20	1.4461e-01	1.0775e-02	9.1373e-02	3.8081e-04
50	8.4193e-03	5.7549e-03	-9.8638e-03	8.7065e-04
100	-3.3910e-04	-1.1624e-03	7.8148e-04	0.0017
200	1.2346e-05	6.1345e-05	-5.7199e-05	0.0034
1000	-2.7158e-08	7.3125e-08	-8.3242e-08	0.0163
10000	2.3910e-12	1.6404e-11	2.3622e-12	0.1623
100000	-1.4211e-14	4.4853e-14	-1.1102e-15	1.6241

Table 12: For a Merton process with T=1

N	K=80	K=100	K=120	CPU time (in seconds)
20	-7.4035e+01	-1.0175e+02	-3.2632e+01	3.8538e-04
50	-1.0907	1.4149	-9.9021e-01	8.8005e-04
100	-2.9049e-05	-2.3883e-05	3.5686e-05	0.0017
160	1.7764e-14	1.7764e-14	-4.9738e-14	0.0027
200	0	0	0	0.0033

Table 13: For a Merton process with T=0.2

N	K=80	K=100	K=120	CPU time (in seconds)
20	-3.8377	-6.3830	4.8290	3.7617e-04
50	-1.8180e-01	3.8766e-01	-2.5272e-01	8.8550e-04
100	4.1351e-04	-5.6452e-04	-1.3386e-04	0.0018
160	2.8622e-09	-7.1283e-09	-5.9122e-09	0.0028
200	-2.4158e-13	-2.9576e-13	1.4522e-13	0.0034
1000	0	0	0	0.0166

Appendix D : Results of the calibration using $w_i = \frac{1}{P_i^2}$

Table 14: Calibration results for the VG model and T=1

Method used	σ	θ	ν	Error	CPU time (in seconds)
CM Lin Int	0.1570	-0.1775	0.8151	1.0590e-03	13.2
CM Cub Spl Int	0.1571	-0.1776	0.8138	1.0641e-03	108.4
cos	0.1567	-0.1782	0.8084	1.1233e-03	15.0

Table 15: Calibration results for the VG model and T=0.75

Method used	σ	θ	ν	Error	CPU time (in seconds)
CM Lin Int	0.1540	-0.1896	0.6314	1.6521e-04	10.1
CM Cub Spl Int	0.1540	-0.1897	0.6302	1.7424e-04	101.6
cos	0.1540	-0.1897	0.6316	1.6802e-04	15.5

Table 16: Calibration results for the VG model and T=0.5

Method used	σ	θ	ν	Error	CPU time (in seconds)
CM Lin Int	0.1468	-0.2108	0.4426	7.2807e-04	10.7
CM Cub Spl Int	0.1469	-0.2107	0.4419	7.1659e-04	111.0
cos	0.1473	-0.2097	0.4466	6.8680e-04	16.2

Table 17: Calibration results for the Merton model and T=1

Method used	σ	λ	μ	δ	Error	CPU time (in seconds)
CM Lin Int	0.0916	0.5157	-0.2403	0.1350	5.1847e-04	33.2
CM Cub Spl Int	0.0967	0.3483	-0.3216	0.0567	9.8092e-04	144.0
cos	0.0919	0.5075	-0.2433	0.1339	5.1587e-04	26.7

Table 18: Calibration results for the Merton model and T=0.75

Method used	σ	λ	μ	δ	Error	CPU time (in seconds)
CM Lin Int	0.0918	0.4986	-0.2543	0.0018	5.4761e-05	19.8
CM Cub Spl Int	0.1102	0.0644	-5.2659	0.2068	0.0068	150.2
cos	0.0864	0.7705	-0.1803	0.1196	5.0738e-05	22.8

Table 19: Calibration results for the Merton model and T=0.5

Method used	σ	λ	μ	δ	Error	CPU time (in seconds)
CM Lin Int	0.0859	0.9241	-0.1603	0.0946	5.9385e-04	23.1
CM Cub Spl Int	0.1141	0.0576	-5.3449	0.2022	0.0365	188.7
cos	0.0861	0.9172	-0.1612	0.0943	5.8101e-04	27.0

Appendix E : Results of the calibration using $w_i = \dfrac{1}{\text{vega}^2}$

Table 20: Calibration results for the VG model, T=1 and vega weights

Method used	σ	θ	ν	Error	CPU time (in seconds)
CM Lin Int	0.1293	-0.2195	0.7030	2.2074e-04	5.2
CM Cub Spl Int	0.1217	-0.2290	0.6781	2.4943e-04	52.6
cos	0.1294	-0.2196	0.7026	2.1975e-04	5.6

Table 21: Calibration results for the VG model, T=0.75 and vega weights

Method used	σ	θ	ν	Error	CPU time (in seconds)
CM Lin Int	0.1191	-0.2601	0.4322	4.1231e-05	5.0
CM Cub Spl Int	0.1581	-0.1811	0.6923	2.0978e-06	61.0
cos	0.1192	-0.2601	0.4320	4.1596e-05	6.2

Table 22: Calibration results for the VG model, T=0.5 and vega weights

Method used	σ	θ	ν	Error	CPU time (in seconds)
CM Lin Int	0.1583	-0.1810	0.5826	3.9179e-05	5.3
CM Cub Spl Int	0.1579	-0.1820	0.5770	3.9198e-05	47.3
cos	0.1580	-0.1818	0.5791	3.7934e-05	7.6

Table 23: Calibration results for the Merton model, T=1 and vega weights

Method used	σ	λ	μ	δ	Error	CPU time (in seconds)
CM Lin Int	0.0985	0.4175	-0.2677	0.1548	2.7597e-05	22.1
CM Cub Spl Int	0.0975	0.4386	-0.2575	0.1595	2.8019e-05	61.5
cos	0.1000	0.3885	-0.2830	0.1485	2.9005e-05	12.3

Table 24: Calibration results for the Merton model, T=0.75 and vega weights

Method used	σ	λ	μ	δ	Error	CPU time (in seconds)
CM Lin Int	0.1053	0.2875	-0.3587	0.1629	3.9413e-05	7.7
CM Cub Spl Int	0.1067	0.2372	-0.4165	0.0935	4.5073e-05	96.6
cos	0.0994	0.3682	-0.3038	0.0016	1.2374e-05	5.8

Table 25: Calibration results for the Merton model, T=0.5 and vega weights

Method used	σ	λ	μ	δ	Error	CPU time (in seconds)
CM Lin Int	0.0964	0.6539	-0.1903	0.1140	4.7446e-05	18.7
CM Cub Spl Int	0.0965	0.6429	-0.1938	0.1114	4.7586e-05	121.0
cos	0.0959	0.6708	-0.1870	0.1150	4.6945e-05	12.4

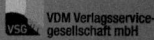

Printed by Books on Demand GmbH, Norderstedt / Germany